Life Cycles

by Peter J. Weber

Contents

Introduction: Growing and Changing 4

Chapter 1

Big Idea Question

How Do Plants Grow and Change? 6
   A Life Cycle Begins 8
   Flowering Plants 12

Chapter 2

Big Idea Question

How Do Humans and Animals Grow and Change? 14
   How Humans Grow and Change 16
   How Humans Are Alike 18
   How Animals Grow and Change 20

Conclusion: Life Cycles on Planet Earth 28

Glossary 30
Index 32

Introduction

Growing and Changing

Plants

Next Generation Sunshine State Standards
SC.2.L.16.1 Observe and describe major stages in the life cycles of plants and animals, including beans and butterflies.

The world is filled with living things. Plants, animals, and humans are living things. All living things grow and change. All living things eventually die.

Animals

Humans

Chapter 1

Big Idea Question

How Do Plants Grow and Change?

Next Generation Sunshine State Standards
SC.2.L.16.1 Observe and describe major stages in the life cycles of plants and animals, including beans and butterflies.

All plants have a **life cycle**. During its life cycle, a plant can grow, change, and make more plants like itself. The plant dies at the end of its life cycle.

A Life Cycle Begins

Not all plants start their life cycles in the same way. Some plants, such as trees, start as seeds.

seed

seed

Other plants start their life cycles as bulbs or cuttings. Still others start their life cycles as spores.

bulb

cutting

spore cases

The kind of seed you plant is the kind of plant that will grow. For example, a bean seed can only **germinate,** or grow, into a bean plant. A young bean plant is called a bean **seedling.**

Beans grow from the flowers. Seeds are inside each bean.

Life Cycle of a Bean Plant

The seed germinates in the soil, and a seedling begins to grow.

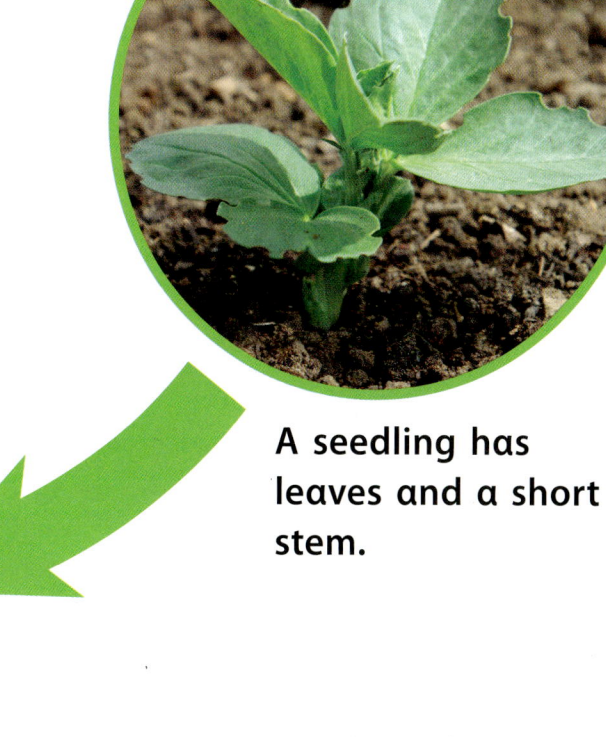

A seedling has leaves and a short stem.

The plant grows taller, and flowers form.

Flowering Plants

Flowering plants, such as a pumpkin, grow flowers that make seeds and fruit.

A pumpkin fruit takes about 160 days to grow from a seed.

This adult pumpkin is an orange fruit. The inside is full of seeds.

Life Cycle of a Pumpkin Plant

A pumpkin seed is white.

A seedling begins to grow in the soil.

A tiny green pumpkin starts to grow at the bottom of the flower.

Chapter 2

Big Idea Question

How Do Humans and Animals Grow and Change?

Next Generation Sunshine State Standards

SC.2.L.16.1 Observe and describe major stages in the life cycles of plants and animals, including beans and butterflies.

SC.2.L.14.1 Distinguish human body parts (brain, heart, lungs, stomach, muscles, and skeleton) and their basic functions.

Humans and animals grow and change during their life cycles. They are born or hatched. They can make more humans and animals like themselves. They also eventually die.

How Humans Grow and Change

You were once a baby. Now you are a child. Soon you will be a teenager. Then you will grow up to be an adult.

adult

baby

Do you look like someone in your family? As you grow and change, you may look more like others in your family. But adults in a family don't always look the same.

teenager

child

How Humans Are Alike

Humans have some things in common. All humans have parts that are the same.

NATIONAL GEOGRAPHIC Fun Facts

Human Body Systems

Digestive (stomach) Your stomach is like a stretchy sack, or bag. A newborn baby's stomach is only as big as a large marble.

Nervous (brain) The brain is part of a message system in your body. It takes in information from your eyes, ears, nose, tongue, and skin.

Muscular (muscles) It takes 72 different muscles to talk. It takes 17 muscles to smile and 43 to frown.

Skeletal (bones) More than half of the bones in your body are in your feet and hands.

Respiratory (lungs) You breathe about 10 million times per year.

Circulatory (heart) Your heart will beat 3 billion times in your lifetime.

- brain
- muscles
- heart
- bones
- lungs
- stomach

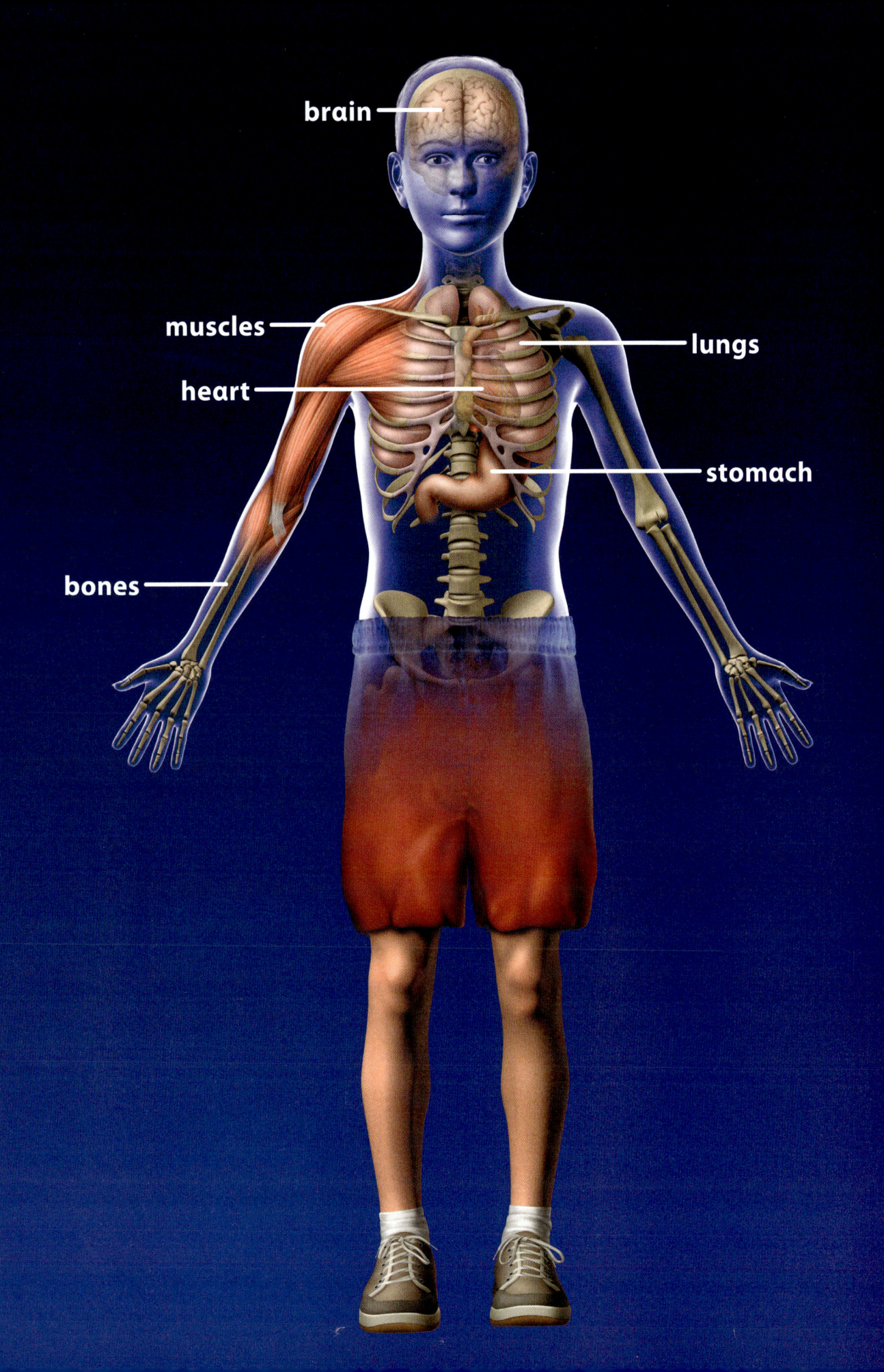

19

How Animals Grow and Change

Animals begin their life cycles in different ways. Most **fish** and **reptiles** hatch from eggs.

A dog is a **mammal**. Most mammals give birth to live young.

This young puppy looks a lot like its parent.

A chicken is a **bird**. What does a chicken's life cycle look like?

A female chicken is called a hen. Some hens lay more than 100 eggs each year.

A Chicken Life Cycle

An egg needs warmth to grow and develop. If the egg gets cracked, a chick will not grow.

After about 21 days, a chick hatches from its egg.

23

A frog is an **amphibian**. What does a frog's life cycle look like?

An adult frog has four legs and no tail. Female frogs can lay eggs.

A Frog Life Cycle

A frog begins its life as an egg in water.

A tadpole hatches from an egg. A tadpole has a tail and lives in water.

A froglet, or young frog, has a tail, too. But it also has legs. It lives mostly on land.

25

A butterfly is an **insect**. Some insects have life cycles with three stages. A butterfly has a four-stage life cycle. What does a monarch butterfly's life cycle look like?

After about 14 days, a butterfly comes out of the chrysalis.

A Monarch Butterfly Life Cycle

An egg is ready to hatch in about one week.

A caterpillar, or larva, can move, eat, and grow.

A caterpillar becomes a pupa. It forms a chrysalis, or covering. It doesn't move or eat.

Conclusion

Life Cycles on Planet Earth

All plants have life cycles. Some plants begin their life cycles as seeds. They grow and change during the stages of their life cycles.

Humans and animals have life cycles, too. They are born live or hatch from an egg. They grow and change during the different stages of their life cycles.

Glossary

amphibian (page 24)

An **amphibian** is an animal that usually starts its life cycle as an egg.

A frog is an **amphibian.**

bird (page 22)

A **bird** is an animal that starts its life cycle as an egg.

A chicken is a **bird.**

fish (page 20)

A **fish** is an animal that usually starts its life cycle as an egg.

A trout is a kind of **fish.**

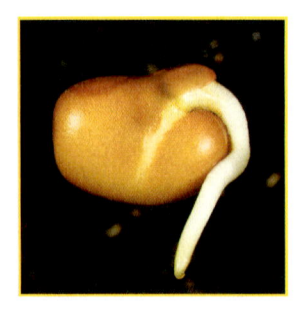

germinate (page 10)

When seeds **germinate,** they begin to grow.

When bean seeds **germinate,** bean plants begin to grow.

insect (page 26)

An **insect** is an animal that usually starts its life cycle as an egg.

A butterfly is an **insect** with six legs and a pair of wings.

life cycle (page 7)

A **life cycle** is the way a living thing grows, changes, makes more living things like itself, and dies.

All plants and animals have **life cycles.**

mammal (page 21)

A **mammal** is an animal. Most mammals give birth to live young.

A dog is a **mammal.**

reptile (page 20)

A **reptile** is an animal that usually starts its life cycle as an egg.

A grass snake is a **reptile.**

seedling (page 10)

A **seedling** is a young plant.

This **seedling** is a young bean plant.

Index

amphibian 24, 30

bean 10–11

bird 22, 30

fish 20, 30

germinate 10, 11, 30

insect 26, 30

life cycle 7–9, 11, 13, 15, 20, 22, 23, 24–28, 31

mammal 21, 31

pumpkin 12–13

reptile 20, 31

seed 8, 10–13, 28

seedling 10, 11, 13, 31

Copyright © 2011 The Hampton-Brown Company, Inc., a wholly owned subsidiary of The National Geographic Society, publishing under the imprints National Geographic School Publishing and Hampton-Brown.

All rights reserved. No part of this book may be reproduced or transmitted in any form or by any means, electronic or mechanical, including photocopying, recording, or by an information storage and retrieval system, without permission in writing from the Publisher.

National Geographic and the Yellow Border are registered trademarks of the National Geographic Society.

National Geographic School Publishing
Hampton-Brown
www.NGSP.com

Printed in the USA.
RR Donnelley, Jefferson City, MO

ISBN: 978-0-7362-7557-6

11 12 13 14 15 16 17

10 9 8 7 6 5 4